This Book Belongs To:

_____

Thank you! Please rate and leave feedback. We appreciate your input.

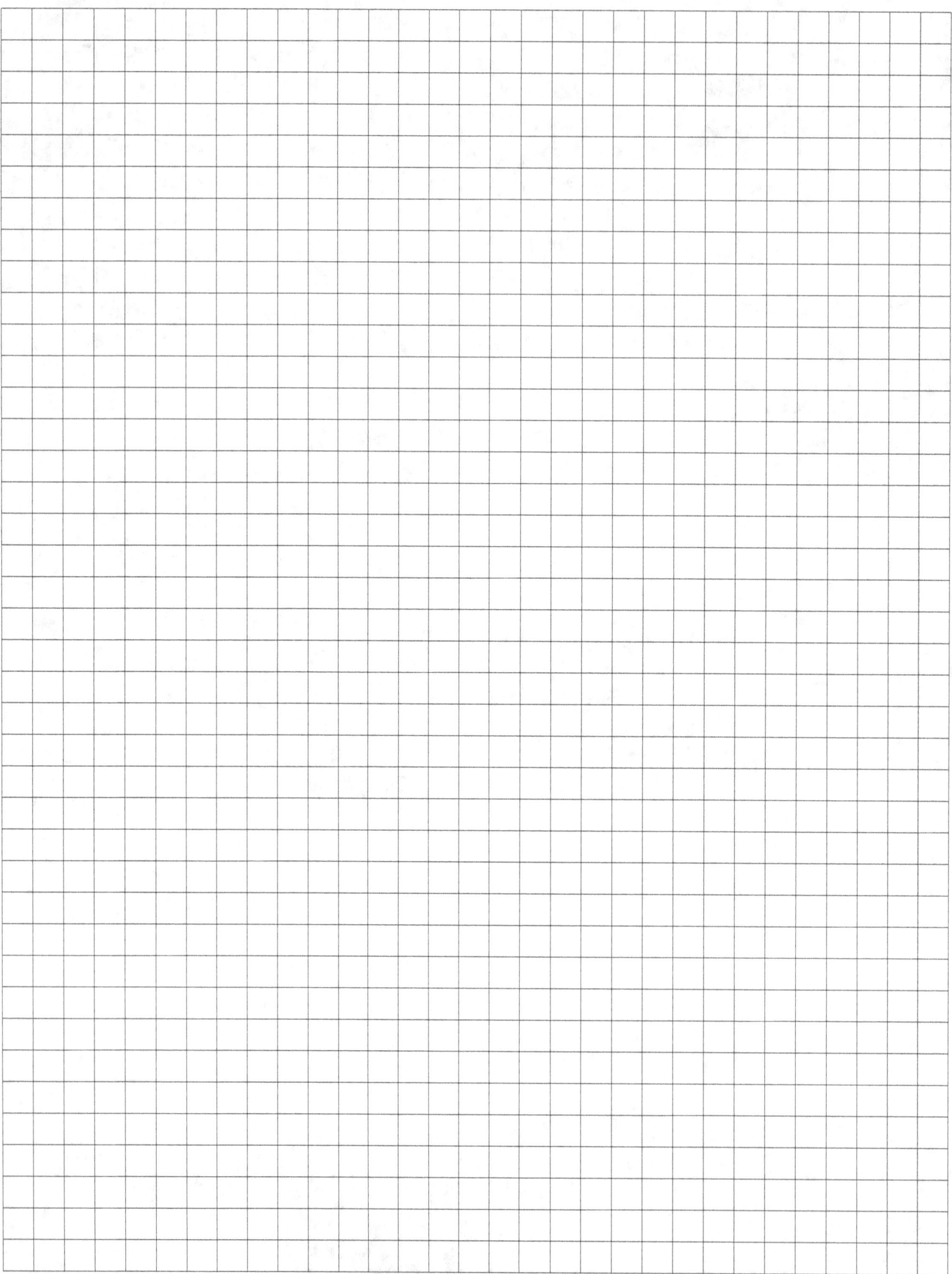

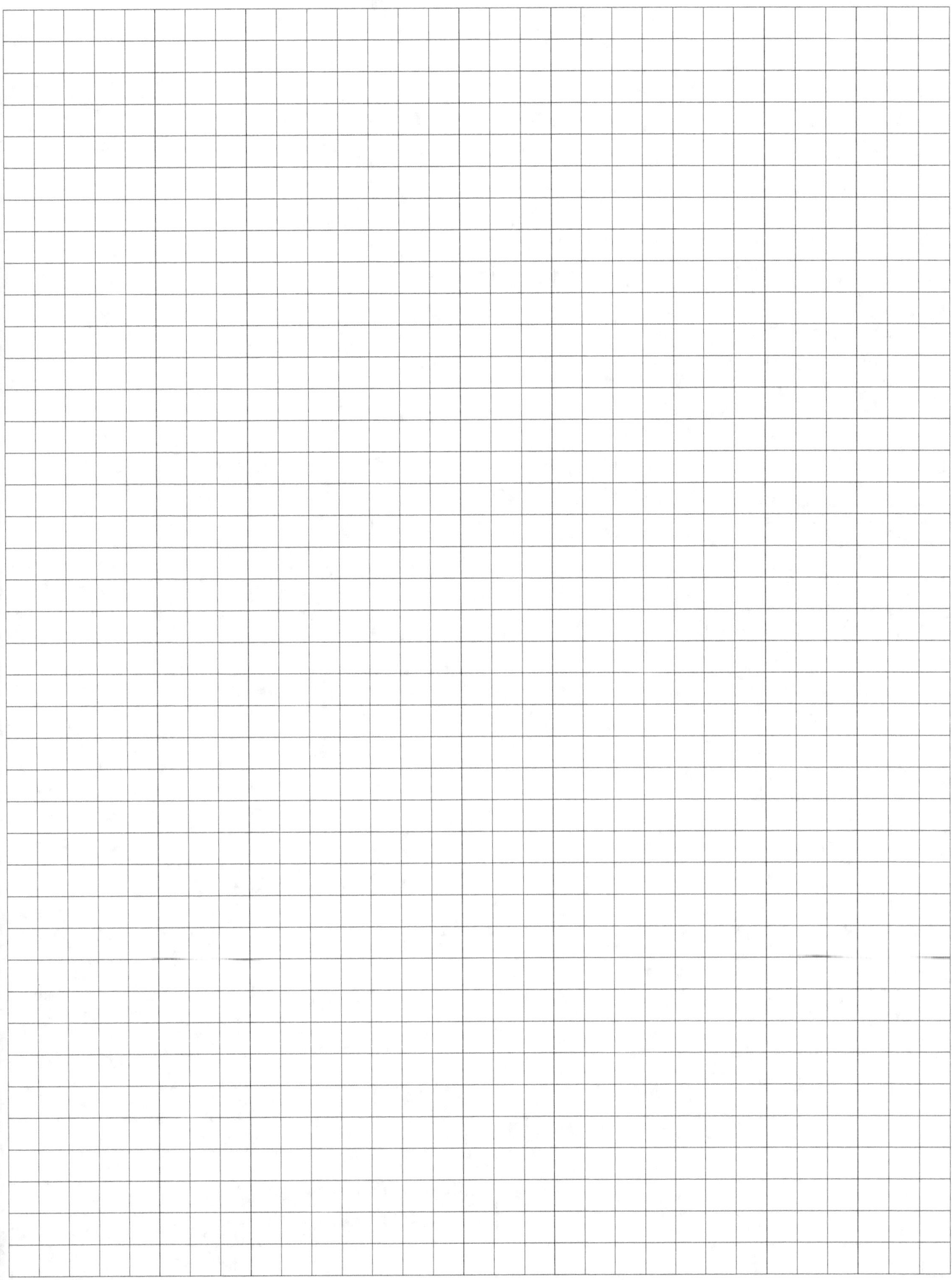

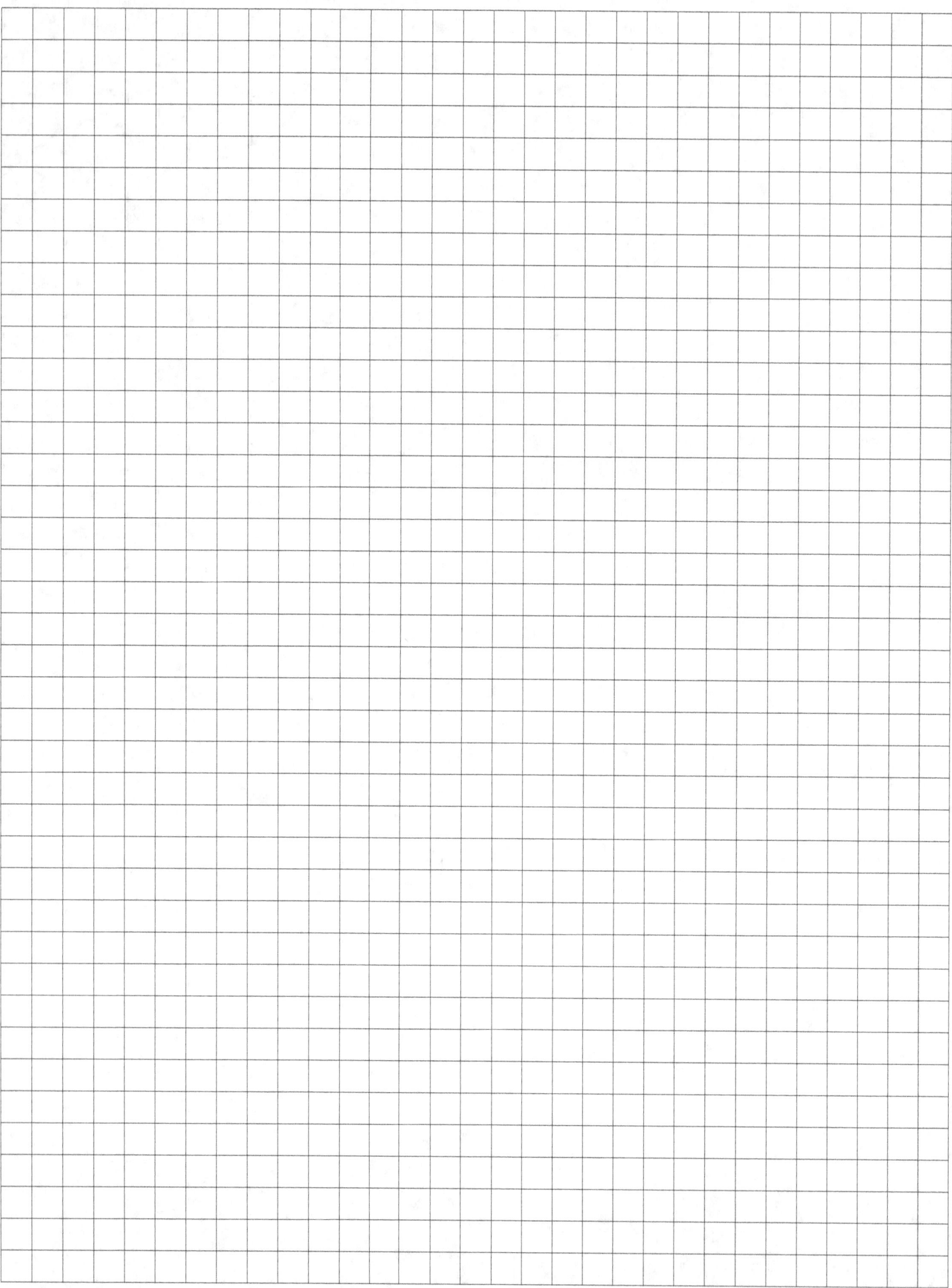

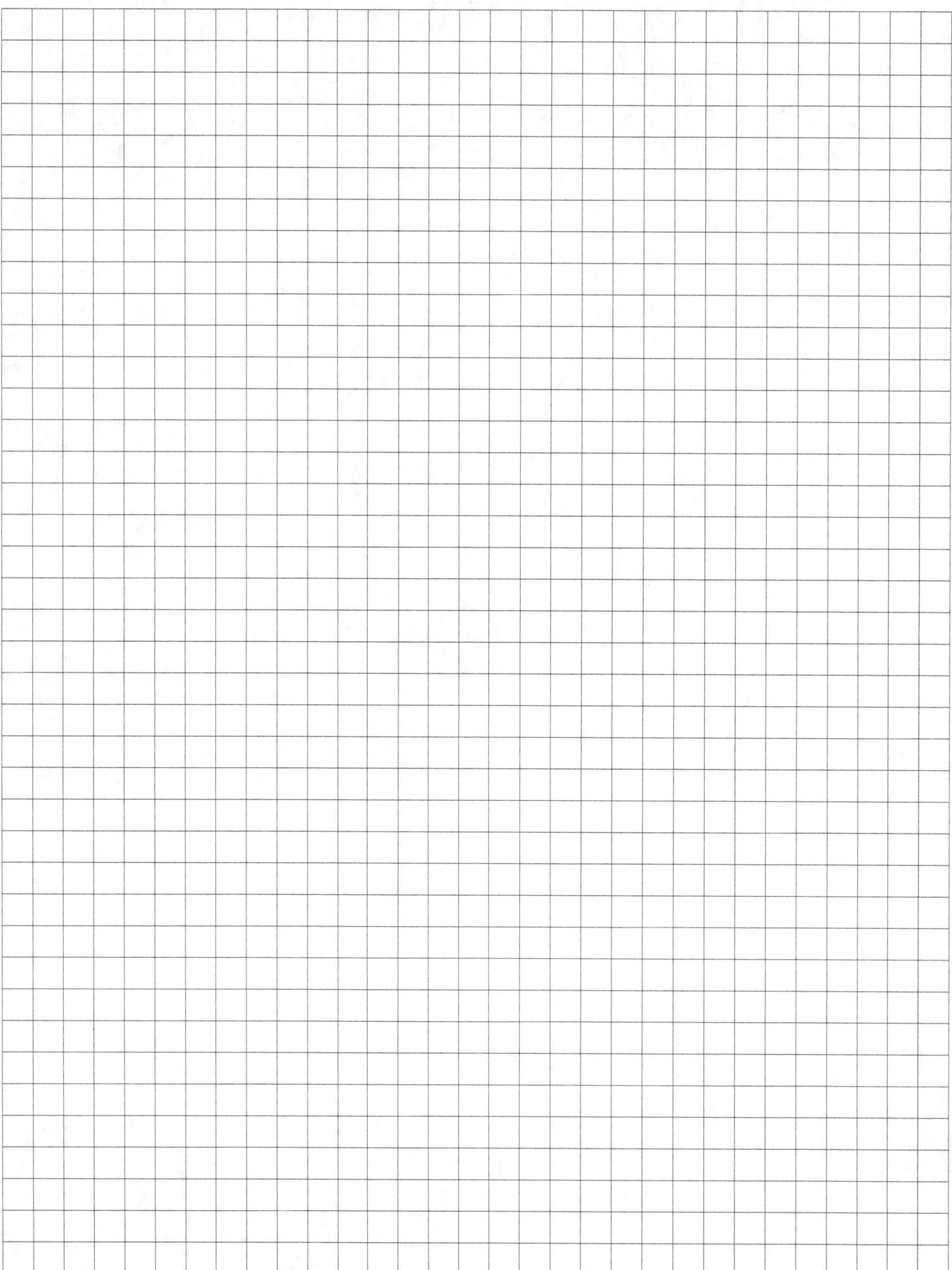

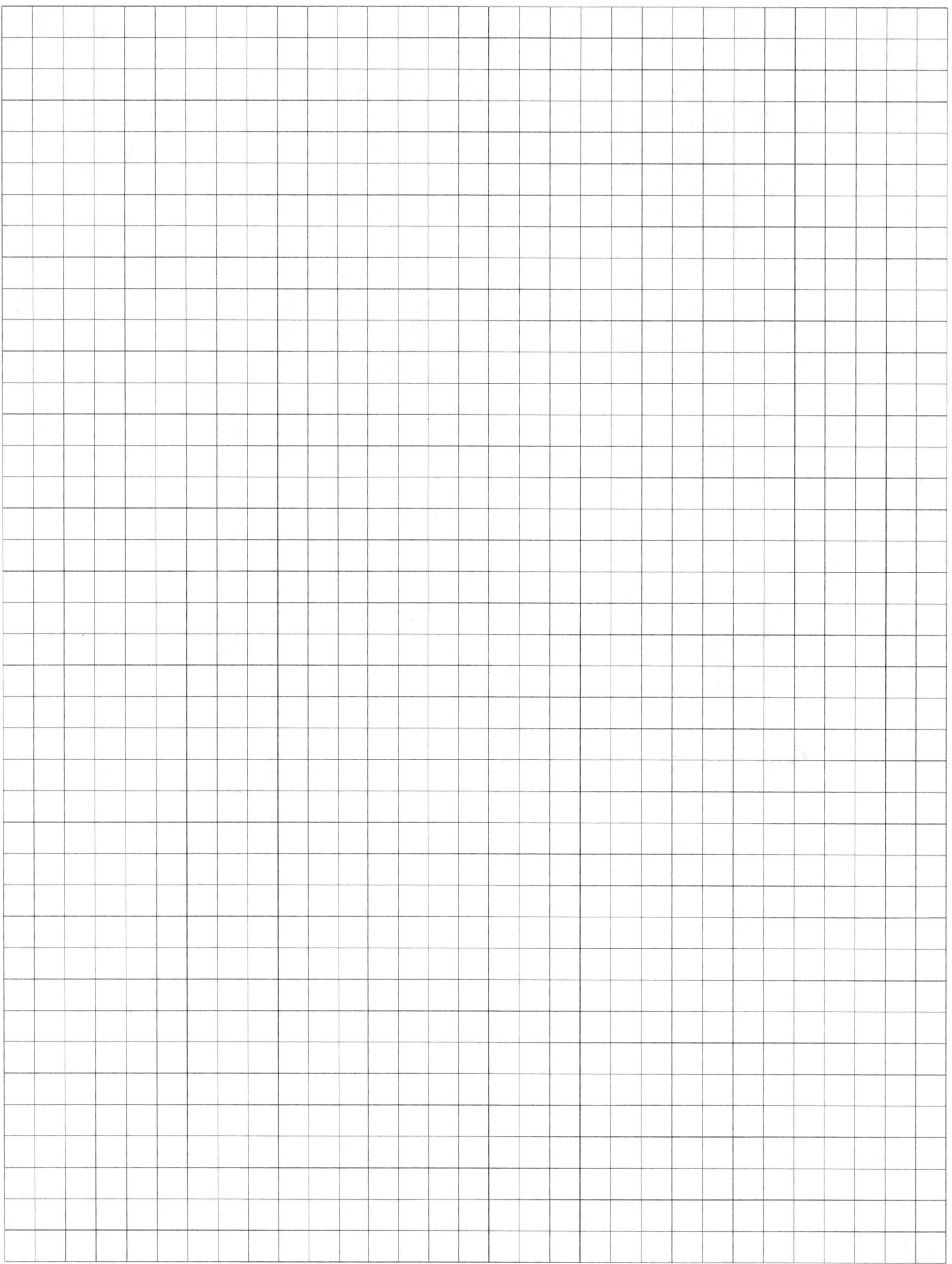

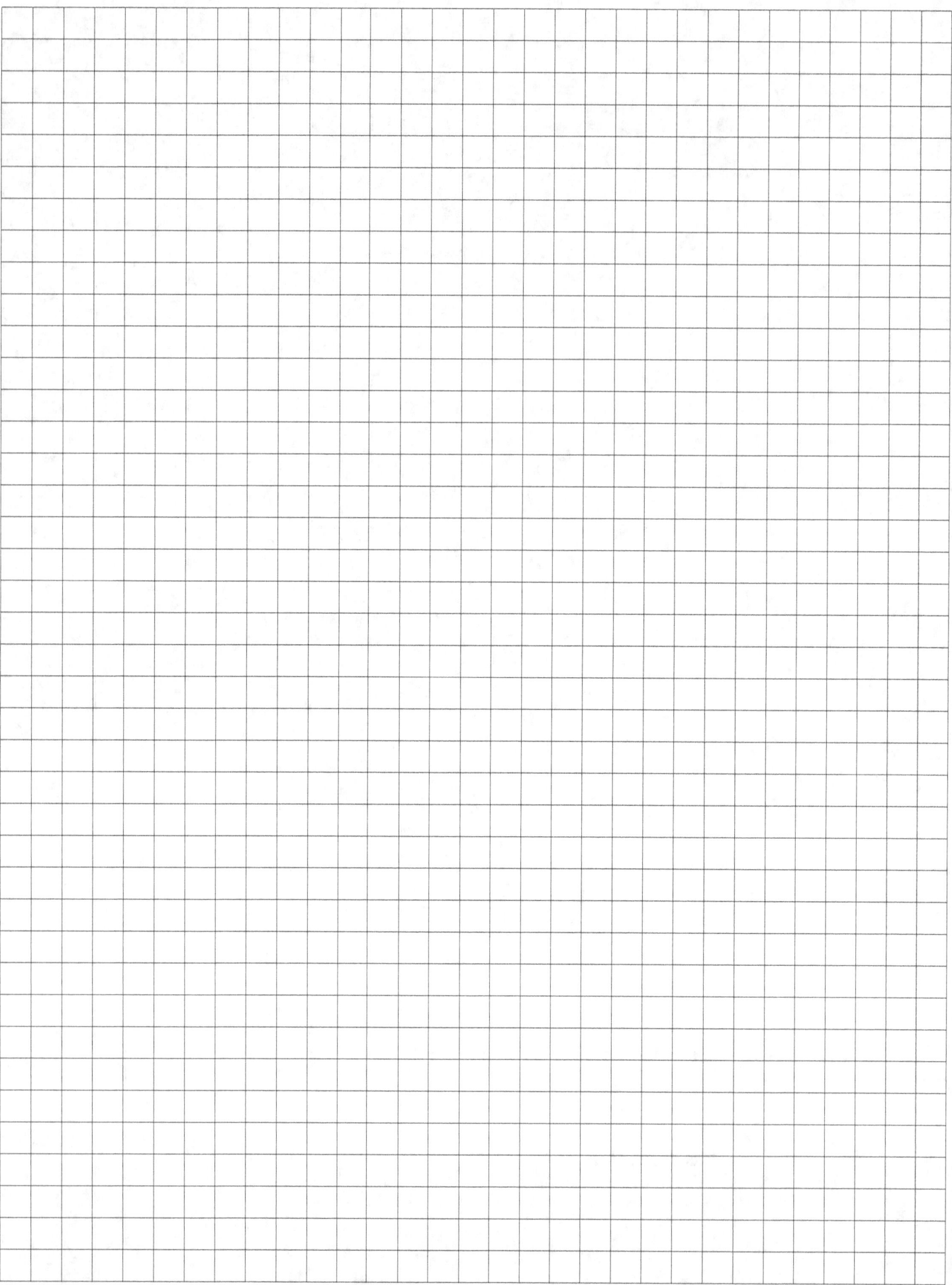

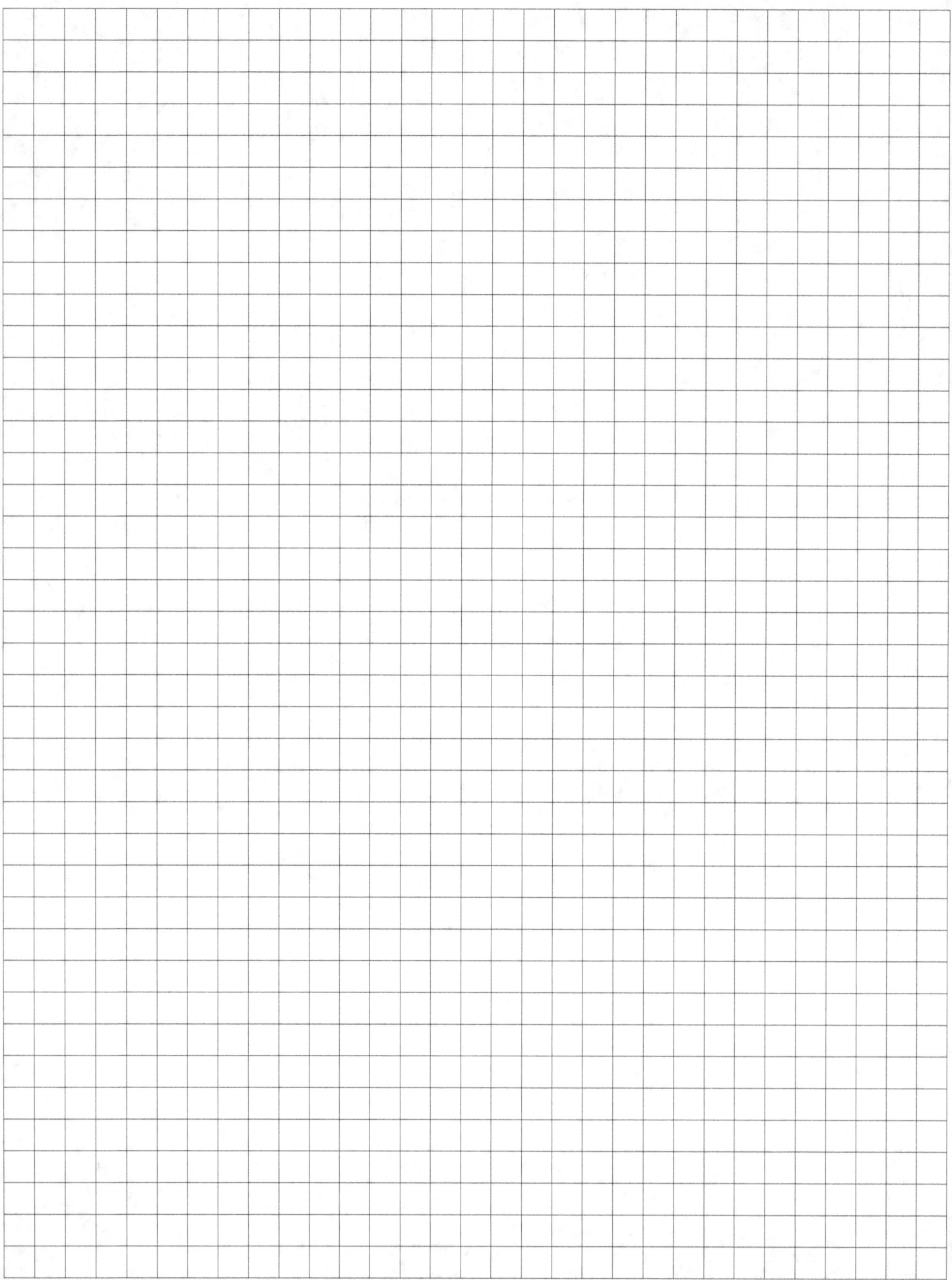

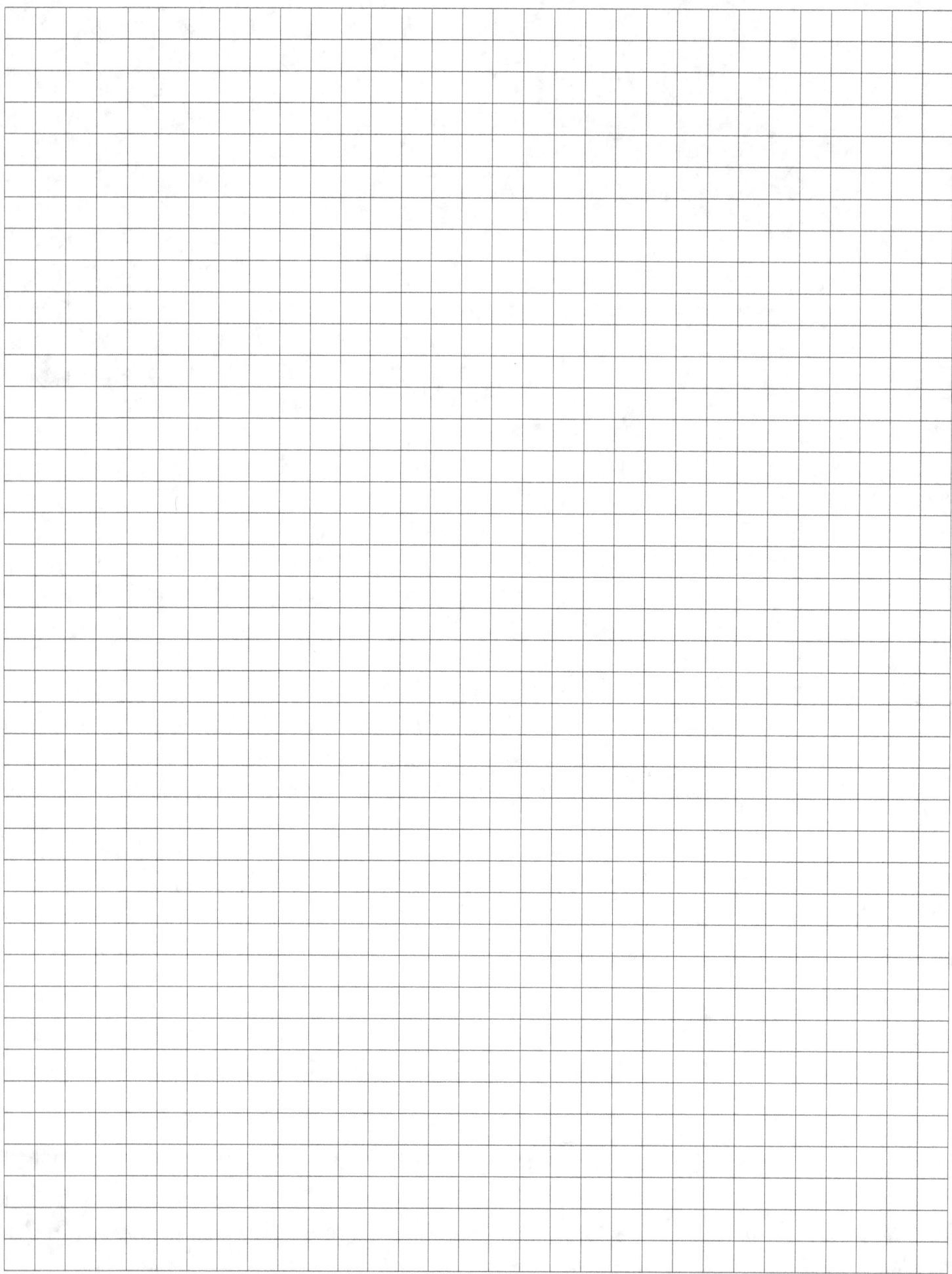

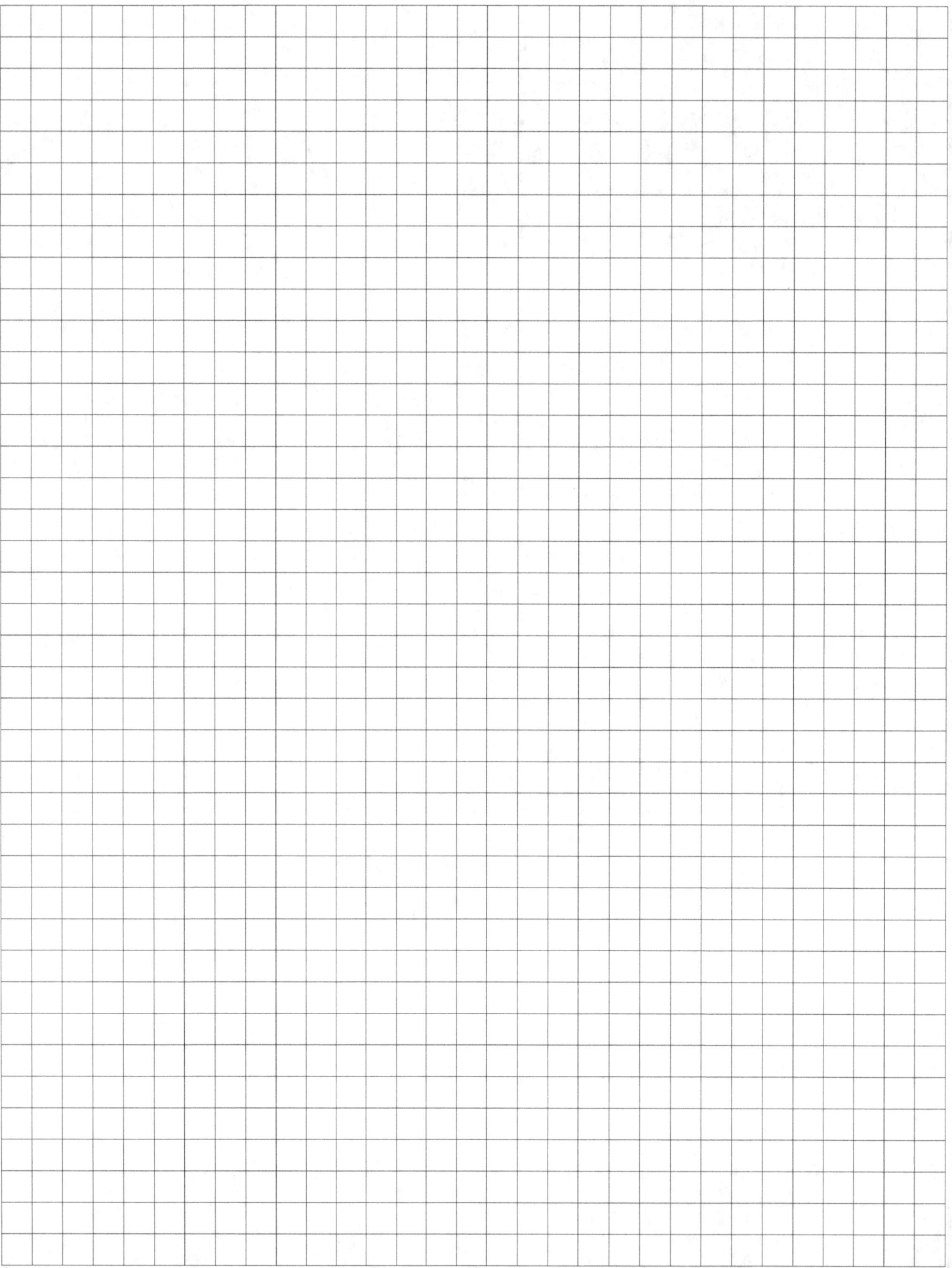

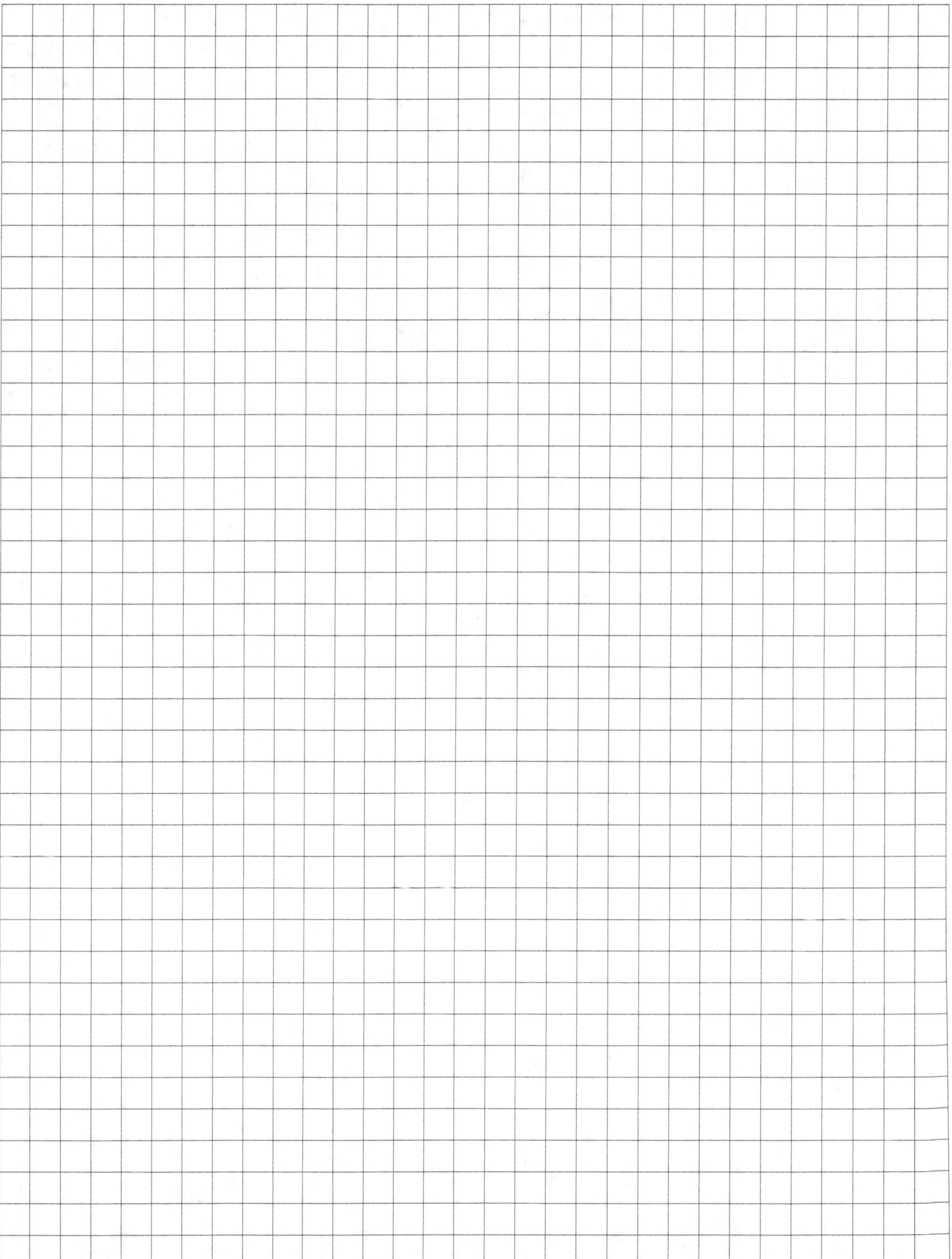

www.ingramcontent.com/pod-product-compliance
Lightning Source LLC
Chambersburg PA
CBHW080548220526
45466CB00010B/3070